MS Environment Library No 8

THE EL NIÑO PHENOMENON

The UNEP/GEMS Environment Library

Earthwatch was the name given in 1972 by the UN Conference on the Human Environment to the assessment activities included in its action plan. Under the plan, each UN agency monitors and assesses those aspects of the environment that fall within its mandate. This Global Environment Monitoring System (GEMS) was formally created two years later, in 1974, and the system is coordinated by the United Nations Environment Programme (UNEP) and its partner agencies through a Programme Activity Centre at UNEP's Nairobi headquarters.

GEMS now has more than a decade of solid achievement behind it. In that time, it has helped make major environmental assessments of such things as the impact of global warming, the pollution of urban air and freshwater resources, the rate of degradation of tropical forests and the numbers of threatened species—including the African elephant—in the world.

As is proper, the results of these assessments have been regularly published as technical documents. Many are now also published, in a form that can be easily understood by those without technical qualifications, in the UNEP/GEMS Environment Library.

This is the eighth volume in the series, and deals with a subject of extraordinary interest: a change in weather patterns in the Pacific Ocean that has been known for centuries, alters the direction of ocean currents, and which is almost certainly linked with unusual and sometimes violent changes in weather extending over about one-quarter of the Earth's surface.

Though the phenomenon called El Niño is ancient, our knowledge of its global effects is recent—dating only from the early 1970s, following the collapse of Peru's fishing industry and a particularly heavy El Niño in 1972–73.

The scientific study of El Niños is now beginning to reveal the full extent of the tragic social and economic impact that these events can cause. It also offers hope that, as we learn to predict their occurrence, we will be able to mitigate their consequences.

Michael D. Gwynne

Michael D. Gwynne
Director
Global Environment Monitoring System

Contents

Foreword	3
Overview	4
The scientific background	6
El Niño and conditions in the Pacific	8
El Niño and the global climate system	11
El Niño triggers	13
Teleconnections	14
The effects of El Niño	17
El Niño in South America	18
Wildlife disappears from Pacific Islands	25
Drought in Australia	26
Drought and forest fires in Indonesia	27
El Niño and Indian monsoons	28
El Niño in China	29
Drought in southern and eastern Africa	30
Storms over North America	32
Implications for policy	34
Sources	36

The link between El Niño and extreme weather conditions has made headlines around the world.

Foreword

Places as far apart as Australia and North America, Peru and India are affected by recurrent climatic anomalies known as El Niño. Originating in the Pacific region, El Niño effects a reversal in the direction of winds and ocean currents, and changes in ocean temperature between Indonesia and the Pacific coast of South America. These climatic changes regularly result in torrential rain and floods in the Pacific coastal countries of South America. Further from the epicentre of these disturbances, El Niño's climatic effects are less predictable, but equally destructive.

There is strong evidence to link extremes of temperature, drought, floods and cyclones around the world to El Niño events. As well as killing people and livestock, and destroying buildings and roads, such events can also change the environment to an extent where it can no longer support local communities. Societies and economies are often left in disarray, with the poor suffering the most hardship.

In order to reduce the destruction and human misery that follow in the wake of El Niño, it is important for us to understand as much as possible about this complex phenomenon. Much research is needed into El Niño itself, so that its occurrence, life-cycle and effects on the global climate system can be predicted.

UNEP is working to promote global awareness of El Niño and to assist information exchange on the subject. As part of its World Climate Impact and Responses Programme, UNEP organized a workshop in 1985 for experts from all disciplines to exchange information on the social and economic impacts of the abnormal weather conditions that occurred during the 1982–83 El Niño. Another workshop was held in Bangkok in 1988 to discuss new research findings and establish plans for further research over the next few years.

UNEP has published material on El Niño that is available to the public and to policy makers. Although it cannot alter the course of El Niño, such information can help governments prepare for El Niño events and thereby reduce their devastating impact. If governments do not act, environmental and economic disasters on a global scale could be the result of the next major El Niño event.

Mostafa K. Tolba

Mostafa K. Tolba
Executive Director
United Nations Environment Programme

Overview

El Niño, or the Christ Child, is the Peruvian name for a weather phenomenon that has been familiar to fishermen along the west coast of South America for centuries. Every year around Christmas time, the fishermen notice that their fishing catch drops for a few months and then returns to its normal level. This reduction in yield is a consequence of El Niño, a seasonal change in weather patterns over the Pacific Ocean. These climatic changes reverse the usual east-west direction of the Pacific currents and, as a result, warmer sea-surface temperatures occur from the central Pacific to the South American coast. Cold equatorial upwellings that usually occur in the coastal waters cease, and nutrients are therefore no longer carried from the ocean floor to the sea surface. As a result, plankton die and fish move elsewhere until sea conditions return to normal.

The term 'El Niño' is now more widely used to refer to abnormally intense instances of the same weather phenomena, and these have been linked to extreme weather conditions around the world. Major El Niños occur about three times a decade and, over the past 30 years, have been the subject of much scientific study.

Evidence of El Niño conditions has appeared in written records for more than 400 years, and continuous time-series of sea temperatures, maintained for more than 100 years, also provide evidence of El Niño events. Since the late 1950s there have been seven major events: 1957–58, 1965, 1968–69, 1972–73, 1976–77, 1982–83 and 1986–87. The 1982–83 El Niño was the strongest this century, and was associated with violent storms along the Californian coast, droughts in Australia and cyclones in the Pacific.

Scientific attention was first focused on the phenomenon in the 1950s, at a time when the Peruvian fishing industry was beginning to expand rapidly. The major El Niño of 1957–8 caused fish populations to fall; this resulted in a sharp decrease in the populations of guano-producing birds off the west coast of South America, with numbers falling to about half their previous levels. The birds' guano was used in the fertilizer industry—one of the major industries in Peru—and, as a result of the birds' disappearance, the fertilizer industry declined rapidly and was soon overtaken in importance by fishing.

During the 1960s, a rapid expansion in fisheries placed Peru at the top of the world league of fishing nations. Then,

in 1972–73, the industry collapsed following a particularly heavy El Niño, and did not recover, even when weather patterns returned to normal.

The possible interrelation between El Niño and global weather patterns was first realized at this time and widely reported in the press, especially in relation to simultaneous droughts in the Soviet Union, Africa, Australia and Central America. These events led to a shortage in world food supplies and famine in many areas. Ten years later, the 1982–83 El Niño coincided with the drought that devastated Ethiopia, the Sudan and a number of other African countries.

Relationships between El Niño and other global weather anomalies (known as 'teleconnections') have become a major scientific study. El Niño is known to have a direct impact on the weather over about one-quarter of the Earth's surface; it is therefore reasonable to expect that it also affects the weather over the rest of the globe. Teleconnections are hard to prove, however, since no two El Niño events have had exactly the same climatic characteristics. But there is little doubt that the weather in the western Pacific, Australia and the Indian Ocean is linked to El Niño, although the exact nature of the relationship is far from clear.

Studies have shown that the major El Niños of this century have occurred fairly regularly and are often preceded by a number of warning signs. It is now possible to predict future El Niños, and we need to learn how to prepare for them. While this will not affect the course of El Niño, it could greatly reduce the often tragic social and economic impacts that the phenomenon causes.

Extreme weather conditions in many parts of the world have been attributed to El Niño.

El Niño is known to have a direct impact on the weather over about one-quarter of the Earth's surface

The scientific background

Scientists and crew on board the research vessel *Conrad*, sailing eastwards along the Equator in the Pacific, first noticed that things were not as they should be in September 1982. For one thing, the sea-surface temperature was considerably warmer than it should have been for the time of year. For another, the ship's cook, who relied on the ocean to keep the crew supplied with fresh fish, was getting no bites on the lines slung from the stern.

The measurements that were being made were all anomalous. In fact, they were so unusual that the computer at the Climate Analysis Centre in Washington DC rejected them as the result of faulty instruments. But everything in the area was abnormal. The sea, normally teeming with life, appeared empty; the whales and dolphins that used to delight those on board were nowhere to be seen. The sea birds had also disappeared.

Things came to a head when one of the *Conrad*'s engines developed a problem. Running on only one engine would leave the vessel considerably behind schedule, since the engine had to work against the prevailing east-west current. But all the navigational readings revealed that the vessel was travelling east faster than it should have been; the *Conrad* was, in fact, ahead of schedule. This could only have happened if the ocean currents were flowing from west to east, the reverse of the normal direction.

We now know that the crew of the *Conrad* were witnessing the onset of a climatic phenomenon known as 'El Niño'—the Christ Child—a name coined by Ecuadorian fishermen because the phenomenon normally occurs around Christmas time, and heralds the start of poor fishing conditions. In most years, the

Figure 1a shows normal weather conditions and ocean currents in the Pacific region. The line '0' indicates the level of South American coastal waters under normal conditions.

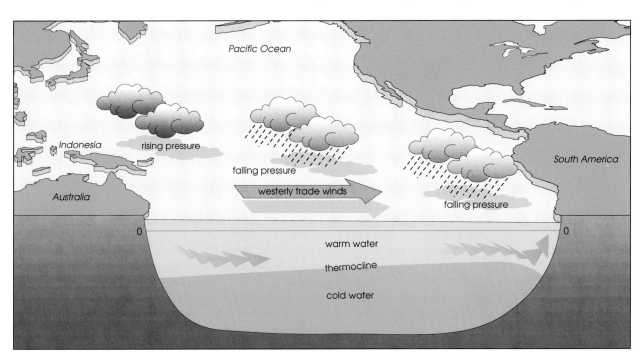

the scientific background

South American fisheries are often affected by El Niños; the Peruvian fishing industry collapsed following the severe El Niño of 1972–73.

changes in weather and sea conditions and the temporary reduction in fish stocks caused by El Niño affect only South America's Pacific coast. But, if El Niño is very intense, the effects spread further afield, and major El Niños are now thought to produce substantial weather changes all over the world.

Figure 1b shows the situation during an El Niño. The line '0' indicates the level of South American coastal waters under normal conditions.

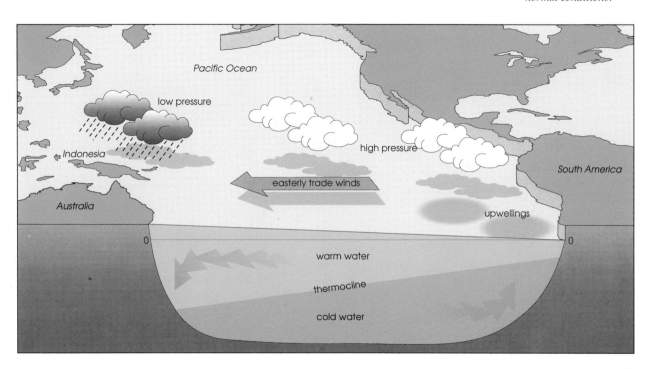

El Niño and conditions in the Pacific

El Niño is a complex phenomenon which we are only now beginning to understand. It embraces seasonal changes in the direction of Pacific winds and ocean currents and abnormally high sea-surface temperatures in the eastern Pacific. These changes normally affect only the Pacific region, but a major El Niño can disrupt weather patterns over much of the globe—sometimes in potentially catastrophic ways—though the causes and interrelation of these events are still poorly understood.

Knowledge of what happens in the Pacific area during an El Niño is much further advanced. Under normal weather conditions, there is a high-pressure system in the tropics in the eastern Pacific (see Figure 1). Over Indonesia, to the west, there is a corresponding low-pressure system. This situation produces easterly trade winds that blow from South America towards Indonesia. These winds tend to push warm surface water towards the west, piling the water up near Indonesia in the form of a small increase (on average, some 40 cm) in the level of the sea surface. Along the equator from the South American coast, the wind causes upwellings of deeper cold water, and the coastal seas there are very cool for the tropics—often as much as 10 °C cooler than Indonesian waters. In August, for example, the temperature of the Pacific surface off Peru may be only 17 °C, while it is 27 °C off Indonesia.

One effect of the build-up of warm water in the west is a corresponding cold undercurrent flowing eastwards, and rising to the surface in 'upwellings' off the South American coast near Peru and Ecuador. The Peruvian fisheries are totally dependent on these upwellings, which carry with them rich nutrient material from the sea-bed on

The importance of upwellings

Under normal weather conditions, the ocean currents off the coast of South America bring deep, cold water to the surface. These cold upwellings contain a copious supply of inorganic nutrients. Plankton breed profusely in the nutrient-rich water and fish that feed off the plankton are abundant. The fish, in turn, support large sea-bird populations, the primary producers for the once flourishing guano fertilizer industry. As the plankton die, they sink down into the deeper, cold water where they decay and replenish the nutrient supply.

The west coast of South America is one of five major upwelling regions in the world. The others are associated with the California Current off the west coast of North America; the Benguela current off the south-west coast of Africa; the Canary Current off the north-west coast of Africa, and the Somali Current off the Horn of Africa. Fish thrive in these coastal waters, and were the basis of huge fishing industries in California, Namibia and Peru.

During an El Niño, sea-surface temperatures along the eastern Pacific coast of South America rise, nutrient-rich cold water is cut off from the surface and fish shoals move elsewhere to feed. During abnormally intense El Niño years, this has seriously affected the catch of the South American, Californian and Namibian fishing fleets (see page 18).

the scientific background

which surface fish feed. The upwellings are driven by a combination of temperature, atmospheric pressure, winds, water currents and the rotation of the Earth. How they interrelate, and the mechanisms that drive them, are largely unknown.

Warm surface waters and cold, deeper water layers in the oceans are separated by a layer of water known as the thermocline, where temperature changes very rapidly. Below the thermocline is cold, stratified water in which temperature falls steadily with depth. Above the thermocline, by contrast, is warmer water which is not in stratified temperature layers.

Off Indonesia, the thermocline is relatively deep—about 200 metres down. However, the effect of the alongshore winds upwelling warm surface water along the South American coast is to position the thermocline in these coastal waters at a depth of as little as 50 metres. The warm surface-water layer off the coast is shallow, allowing upwellings from a depth of few hundred metres to carry nutrients up to the surface waters where shallow-water fish feed on them.

This, then, is the usual situation. What happens during an El Niño? One trigger for the El Niño phenomenon is an eastwards movement of the prevailing low-pressure system over Indonesia, which weakens the high-pressure system near South America. As a result, the easterly trade winds first slacken, then disappear, and finally reverse direction and begin to blow from the west.

With this change in wind direction, the water that has accumulated in the western Pacific now has nothing to hold it in place. The Pacific then behaves like an enormous bath that is suddenly lifted at one end. The

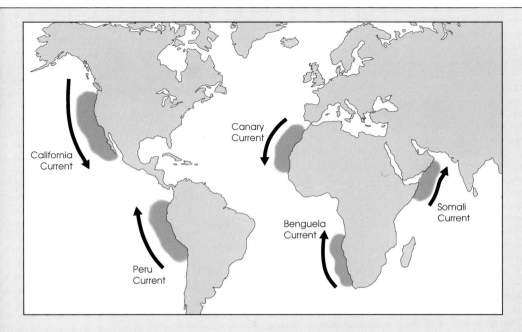

Coastal upwellings are important because they provide abundant nutrients and therefore encourage rich fishing grounds. The currents associated with the world's five major upwelling areas are shown on the map.

The El Niño phenomenon

The sea conditions that normally make the Pacific coast of South America one of the most prolific fishing grounds in the world are disrupted by El Niño, and fish catches fall while El Niño conditions prevail.

warm surface water that has built up in the west slops back towards the east, reversing the direction of the surface current. Sea levels and sea-surface temperature off the South American coast rise, and the thermocline is pushed down to a much deeper level. Upwellings then occur entirely above the thermocline and this cuts off the supply of nutrient-rich water from the ocean floor to the surface.

Deprived of nutrients, plankton in the coastal waters of the south-eastern Pacific begin to die. Their population collapse affects the open-ocean (pelagic) fish that feed on them and that are the basis of the fishing industries of Ecuador, Peru and Chile. The fish shoals move in search of more nutritious waters. Their location is unpredictable and often out of range of the fishing fleets. Catches usually drop while El Niño conditions prevail.

the scientific background

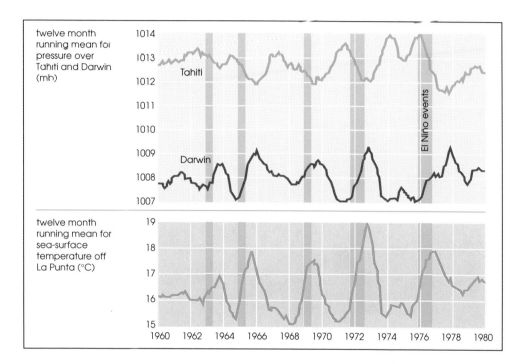

Figure 2 shows that sea-surface temperature off the Pacific coast of South America—an indicator of El Niño—is related to atmospheric pressure over Darwin, Australia and Tahiti. When sea-surface temperature at La Punta in the Pacific is higher than normal, air pressure is correspondingly higher than usual over Darwin and lower than normal over Tahiti.

El Niño and the global climate system

Major El Niño events were recorded in 1877, 1918, 1925, 1940–41, 1957–58, 1965, 1968–69, 1972–73, 1976–77, 1982–83 and 1986–87. The first scientists studying these events observed that the characteristic warming of large areas of the eastern Pacific ocean associated with major El Niños appeared to produce widespread changes in the weather, even over distant continents. However, the climatic mechanisms relating these distant weather phenomenon were not apparent, particularly because no two major El Niño events have been identical.

It was the recognition of the relationship between El Niño and one of the world's better-known atmospheric systems—the Southern Oscillation—that gave scientists a deeper understanding of the global effects of major El Niño events. Scientists had known since the turn of the century that changes in atmospheric pressure over the Indian Ocean (in the west) were always mirrored by opposing changes in pressure over the south-eastern Pacific (in the east): if one was rising the other was falling. This 'see-saw' relationship is known as the Southern Oscillation.

In the late 1960s, scientists realized that these opposing changes in atmospheric pressure over the Indian Ocean and south-eastern Pacific rose and fell in relation to the warming of the eastern Pacific associated with El Niño.

It has now been established that sea temperatures off the coast of South America rise as atmospheric pressure over Darwin, Australia, rises; and that sea temperatures in the eastern Pacific fall when pressure over Darwin falls. Both change in opposition to the rise and fall of pressure over the eastern Pacific (*see Figure 2*). It is precisely because El Niño is related to global weather patterns that an intense El Niño can disrupt weather systems on the other side of the world.

To describe more accurately the interaction of these two phenomena, their names have been joined together as the El Niño/Southern Oscillation or ENSO.

Though no two events are exactly the same, an ENSO typically has four distinct phases: precursor, onset, growth and decay. The precursor starts with an intensification of the prevailing weather patterns. The high atmospheric pressure in the eastern Pacific rises, with a corresponding fall in pressure in the west. The easterly trade winds blow harder, pushing more surface water towards Indonesia, where the sea level rises; and there is an corresponding drop in sea level off South America. The sea-surface temperature increases in the west and drops in the east.

The onset occurs around December. The abnormally intense prevailing conditions suddenly change. Sea-surface temperature drops in the western Pacific and rises in the east. Atmospheric pressure rises over Indonesia and northern Australia, and drops in the eastern Pacific near South America. The trade winds blowing from the east slacken, and then start blowing back across the Pacific from the west. There is increased rainfall in the central Pacific, a normally dry region, and over the coastal regions of Peru, Ecuador and Chile.

The growth phase is a continuation of the onset. Sea-surface temperatures off the coast of South America continue to rise, reaching a maximum in June. The flow of warm water from the western to the eastern Pacific raises sea levels in the east and pushes the thermocline deeper.

These conditions continue to intensify during the year. Warm water continues to be blown across the Pacific until it reaches the American continent, where it is deflected both to the north and south, thus warming the sea throughout the whole of the eastern Equatorial Pacific region. Winds continue to blow from west to east; rainfall decreases dramatically over Indonesia, and falls heavily over the central and eastern Pacific and South American Pacific coast.

Conditions reach a peak about a year after the onset, and then the westerly winds start to weaken, heralding the start of the ENSO decay phase. There is a brief secondary warming of the sea surface off South America, and then, over the next six months, things gradually revert to the normal easterly trade winds and lower sea-surface temperatures. A year and a half after the onset, the Pacific weather has returned to normal.

El Niño triggers

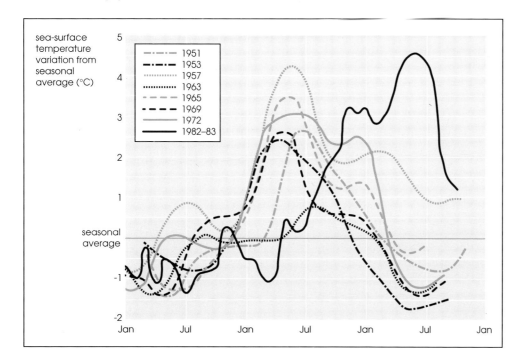

Figure 3 shows higher-than-average sea-surface temperatures in the south-eastern Pacific in eight El Niño years. Such warming usually precedes El Niño events, but in 1982–83 increases in sea-surface temperature occurred near the end of the El Niño.

Current understanding of Pacific conditions suggests that critical factors in triggering an El Niño event are the strength of the trade winds in relation to the amount of water piled up near Indonesia. As long as the trade winds continue to blow hard, they have enough force to hold this excess water in place. But eventually, so much water is accumulated that any random fluctuation in atmospheric circulation that is sufficient to weaken the trade winds for long enough, causes a surge of warm water eastwards across the Pacific.

The sea-surface temperature in the mid-Pacific then rises, warming the atmosphere above and causing rain clouds to develop, winds to change direction, and the low-pressure system over Indonesia to move eastwards over the central Pacific. The scene is now set for an El Niño event. What happens at the other end of the cycle, when an El Niño event ends and normal conditions are restored, is still unknown.

This scenario suggests that an anomalous warming of the eastern Pacific precedes the collapse of the trade winds. This does seem to be the case in most instances. However, in the exceptional 1982–83 El Niño, it appears that the normal sequence of events, which started in the middle of the year instead of December, was reversed: the intense warming of the eastern Pacific occurred near the end of the event (*see Figure 3*), not at the beginning. In addition, both the low-pressure area in the west and the high pressure in the east moved further east than in a normal El Niño event.

This may explain why the 1982–83 event was so intense, and why some of the associated effects in other parts of the world, such as North America, were different from those in other El Niño years.

The El Niño phenomenon

Teleconnections

The relationship between El Niño events and climatic variations in the Equatorial Pacific region is extremely strong and well-documented. It is less easy to prove that weather disturbances far from the Pacific are related to El Niño. Weather anomalies occur all over the globe every year, but some do tend to recur with most or all El Niño events and are referred to as teleconnections.

The Pacific weather system covers about one-quarter of the world's surface area. It is not surprising, therefore, that this weather system should apparently exert a far-reaching influence round the globe. However, the atmospheric mechanisms involved are still largely unknown.

The first investigations of El Niño teleconnections were carried out after the strong El Niño of 1957. One result was the development of a model of the Equatorial atmosphere, comprising three major convection cells over the Pacific, Indian and Atlantic Oceans. In each, warm moist air—characterized by clouds and rain—rises to a height of about 12 km, cools, and then descends as cold, dry air. These 'Walker cells', named after the scientist who first studied the Southern Oscillation in the 1920s, are the engines of atmospheric circulation. Normally, the rising currents of warm air in the cells are situated over Indonesia, Africa and the Amazon; the cold, dry air descends over the Pacific, Indian and Atlantic Oceans, as shown in Figure 4.

During an El Niño, as in 1982–83, the low-pressure zone over Indonesia and its associated warm, moist air move eastwards over the mid-Pacific. The patterns of Walker cells are then greatly changed, as shown in Figure 5. Because this warm air has moved to the east of its usual position,

Figure 4 The Equatorial atmospheric system is the basis of global atmospheric circulation. The system comprises three major convection cells of rising warm, moist air and descending dry, cold air located over the Indian, Pacific and Atlantic Oceans.

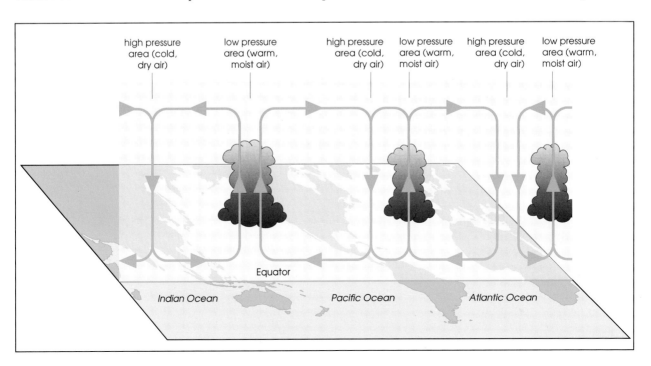

the cold, dry air that usually descends over the Pacific, Indian and Atlantic Oceans is also moved further east than normal, and falls over Australia and Africa.

This significantly alters global rainfall patterns because cold, dry air—associated with low rainfall—has replaced the warm, moisture-bearing air formerly centred over Australia/Indonesia and Africa. This would account for the droughts that occurred in Africa and Australia at the same time as the 1982–83 El Niño. These movements in the location of the Walker cells can also be used to explain why other areas receive exceptionally high rainfall during El Niño events.

For instance, major El Niños usually result in heavy rainfall in western South America: in Ecuador most of the years since 1951 with abnormally heavy precipitation have also been El Niño years.

This is a result of the shift in location of the Walker cells. The increase in sea-surface temperature of the eastern Pacific heats up the air above it, and effectively moves the Walker cell, which usually bring rain to Indonesia to the east and causes rain to fall over the western Andes instead. However, not all El Niño years since 1957 have brought heavy rainfall; in some cases the increase occurred after the El Niño.

Recent studies have confirmed that there are correlations in many other areas of the world between changes in rainfall and El Niño events. Correlations are strong in the following areas: the central Pacific and the south-east of South America have above-average rainfall during El Niño months, as does Equatorial eastern Africa. Rainfall is well below average during El Niño over Papua New Guinea; northern, eastern and central Australia; north-eastern South

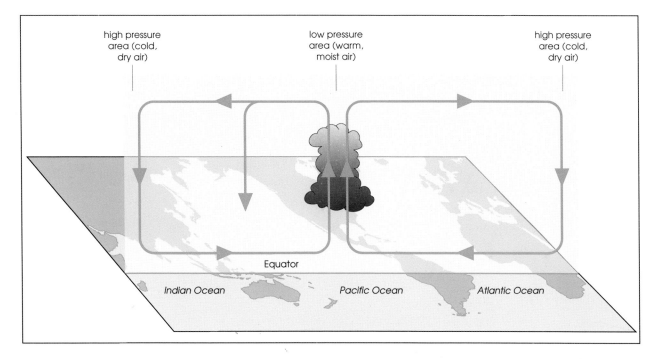

Figure 5 *During an El Niño, changes in atmospheric pressure result in warm, moist air moving eastwards from Indonesia to the South American coast and cold, dry air, formerly over the oceans, moving east over the major land masses.*

The El Niño phenomenon

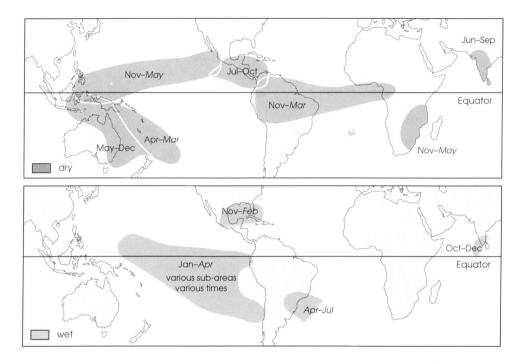

Figure 6 shows typical anomalies in rainfall caused by El Niño. Months in roman type are in the year of the onset of an El Niño, those in italic indicate the following year.

America; and India and south-eastern Africa.

Correlations are weaker for Central America, which tends to be drier in El Niño years. Rainfall in North America does correlate somewhat with El Niño but the correlation is inconsistent: the weather was drier in 1976 and wetter in 1982. In north Africa, southern Europe and the Middle East there are no detectable correlations with rainfall.

Studies on the occurrence of hurricanes throughout this century have now established that they are less common in the western Atlantic during El Niño years and that cyclone activity in south-east Asia tends to move eastwards in El Niño years.

Although the case for teleconnections between El Niño events and disturbed weather conditions is established for many areas of the world, the detailed mechanisms involved are largely unknown. What is known is that the effects produced by El Niño are extremely variable, and that the strongest teleconnections occur in the tropics, and the weakest and most unpredictable in extreme latitudes.

Studies using ocean and atmospheric models and evidence from global palaeo-climate records such as sediments, coral records and tree rings indicate that the frequency and intensity of El Niño events have not varied a great deal since the little ice age (1500–1850). However, the UNEP Working Group on the Socio-Economic Aspects of ENSO at its third meeting in November 1991 considered that global warming could possibly change the frequency and intensity of ENSO. Model results suggest an intensification of climate effects in the tropics with anomalously wet areas becoming wetter and dry areas drier.

The effects of El Niño

Weather, alongside war, is one of the most disruptive forces affecting human society. In any year there are gains and losses in different regions of the globe due to climate, but in El Niño years the losses tend to outweigh the gains heavily. During the severe 1982–83 El Niño, weather-related damage was estimated at US$8.7 billion. Millions of deaths, massive livestock mortality, incalculable damage to wildlife and hardship spread across at least a quarter of the globe. There were floods and droughts in South America, raging bush fires and droughts in Australia, drought and famine in Indonesia, devastating storms and mud slides along the Californian coast, and drought in the Sahel and southern Africa.

Climatic change and abnormal weather can pose a direct threat to people and to wildlife from floods, storms and forest fires. People are also prone to the diseases that result from damage to shelter and reduced food and water supplies.

Other effects are not as immediately apparent, but can cause long-term problems. The delicate balance of eco-systems can be disrupted, for instance, as animals and birds die because changes in temperature or rainfall have reduced their food source. Soil quality can be reduced through wind and rain erosion, especially in drought-struck areas, leading to lower crop yields and, if farming becomes impossible, to the rural population drifting towards towns and cities. On the economic side, disrupted transport and communications can prevent industry from working properly, leading to shortages of commodities, falling incomes and the loss of international markets and foreign exchange for the national economy.

But perhaps the most insidious result of El Niño's devastation is that resources earmarked for projects to provide protection against future drought or downpour have to be diverted to cope with the immediate emergencies. This diversion delays, and often prevents, the completion of the projects, leaving people vulnerable to further disasters.

This section examines the ecological, social and economic effects of El Niño events, particularly those that have occurred since the late 1950s. While countries such as Australia, Chile, Peru, Ecuador and Indonesia are nearly always affected by predictable weather phenomena during El Niño, countries farther from the Pacific region experience less regular effects. In such areas of the world, El Niño events can be associated with any type of extreme weather—drier, wetter, warmer or colder than normal. During the El Niño of 1976–77, for example, the eastern United States suffered bitter winter weather. Supplies of fuel were nearly exhausted in several regions. But, during the 1982–83 El Niño, the same area experienced an unusually mild winter; savings in heating costs were estimated at more than $2 billion.

A greater knowledge of the triggers and life cycle of El Niño events may help us to predict future occurrences and thus minimize both their immediate and long-term effects. Studying the extent of the damage of past El Niños in different countries can reveal what sort of agricultural and economic systems, for instance, are the most susceptible to disruption by climatic anomalies and what policies would effectively reduce this disruption.

> During the severe 1982–83 El Niño … there were floods and droughts in South America, raging bush fires and droughts in Australia, drought and famine in Indonesia, devastating storms and mud slides along the Californian coast, and drought in the Sahel and southern Africa.

The El Niño phenomenon

El Niño in South America

Fertilizer and fishing industries in Chile and Peru

The waters along the coasts of Peru, Ecuador and northern Chile are normally some of the most productive in the world. Nutrient-rich currents rising from the sea-bed support vast populations of microscopic plankton, which in turn provide food for fish and other wildlife. The system is delicately balanced, with dead plankton falling to the sea floor and decaying to provide the nutrients for future upwellings.

During an El Niño, plankton populations drop dramatically, disrupting the food chain and leading to starvation among marine life forms. Over the past 40 years two of the major industries of Chile and Peru—fertilizer and fishing—have been directly affected by these changes.

In the first half of this century, fertilizer made from the droppings of guano sea birds living on off-shore islands was central to the Peruvian economy. The sea birds fed on anchoveta fish. Guano fertilizer was exported in bulk, and used domestically to improve crop yields.

The changes in sea conditions caused by the 1957–58 El Niño led to a massive drop in anchoveta numbers. Millions of guano birds starved to death and their number dropped from 30 million to 16 million. A few years later, Peru began to expand its anchoveta fishing industry. Competition from fishermen for the anchoveta and stresses caused by El Niño caused a further decline in the number of guano birds and ended the dominance of the guano fertilizer industry in Peru.

The fishing industry continued to grow and, by the late 1960s, Peru had become the world's leading fishing nation by

Only a small number of the millions of sea birds that once lived along the Peruvian coast survived the El Niño events and overfishing of the 1950s and 1960s.

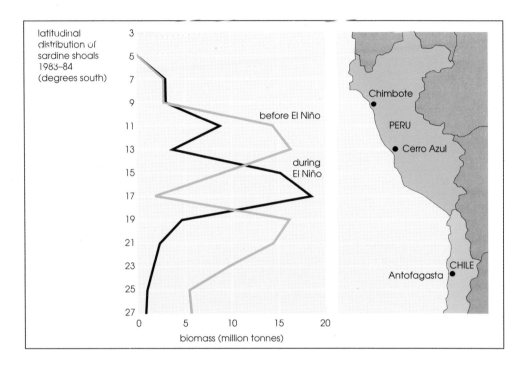

Figure 7 During the 1982–83 El Niño, sardine shoals moved southwards—out of the range of the main Peruvian fisheries in the north of the country—and came within range of the Chilean fishing fleet, increasing its catch significantly.

weight of catch, with anchoveta making up the bulk. By 1970, 1400 modern boats were bringing in 14 million tonnes of anchoveta a year, one-fifth of the world fish catch.

However, the 1972–73 El Niño devastated the Peruvian anchoveta fishing industry and brought the El Niño phenomenon onto the world stage for the first time. During the onset of the 1972–73 El Niño, the warming of the Pacific coastal waters drove large numbers of anchoveta close to the shore—the only place where cold, plankton-laden upwellings were still occurring. The Peruvian fleet took record catches of up to 180 000 tonnes a day, giving the appearance of a bumper year for anchoveta.

In reality, the anchoveta population was close to collapse. The conjunction of overfishing during 1971 and the El Niño onset caused anchoveta numbers to plummet; the annual catch dropped from about 12 million tonnes in 1970 to 2 million tonnes in 1973. Boats stood idle and fishmeal plants were brought to a standstill. The Peruvian government responded by nationalizing the fishing industry. By 1976 there had been no resurgence in the anchoveta populations so the government denationalized the industry as it could no longer afford to subsidise the fishermen and processing workers.

During the 1982–83 El Niño, fish populations moved along the South American coast so that catches in some areas increased, while others dropped. For example, the sardine population off northern Chile became concentrated near the coast during the first half of the year and Chilean catches increased substantially. At the same time, the northern Peruvian sardine population moved out to greater depths and migrated south, as shown in Figure 7, so that sardine catches fell significantly in Peru.

Other species of fish moved around South America's coastal waters in different directions. Jack mackerel were absent from the normal Chilean fishing areas. In Peru, no alteration in jack mackerel populations was noticed, but these fish were caught for the first time in commercial quantities in Ecuador, suggesting that the fish had moved north in search of food.

Overall, the short-term effect of El Niño was to increase the total Chilean fish catch during 1982–83, because larger sardine catches compensated for the drop in jack mackerel. However, total Peruvian and Ecuadorean catches of sardine decreased significantly at this time; Peru suffered a dramatic fall in anchoveta landings from 1.7 million tonnes in 1982 to less than 120 000 tonnes in 1983.

Although Chile gained a short-term benefit from the increased accessibility of sardine, El Niño had long-term detrimental effects on the South American fishing industry. In 1982–83, it seemed that anchoveta population levels had been more severely reduced by the El Niño than those of the sardine. But this reduction was noticed immediately only because anchoveta can be fished after one year, whereas sardine take four to five years to reach commercial fishing weight. In 1982–83, however, sardines lost an average of 20 percent of their body weight in both Chilean and Peruvian waters and did not spawn during the summer of 1983. This decrease in growth and fertility had a delayed, though major, effect on the sardine population and the South American fishing industry.

As well as disrupting commercial fisheries, the 1982–83 El Niño affected Chile's numerous poor, small-scale, shore-based fishermen, whose main catches are sea urchins and false abalones. Widespread mortality in these species led to a moratorium on fishing to allow populations to recover, causing severe hardship to the fishing communities.

However, the warm water did fuel a huge increase in the shrimp population. One Peruvian processing plant at Tumbes that had been taking 5 tonnes of shrimp each week turned out 75 tonnes during the El Niño period.

The El Niño years of 1972–73 and 1982–83 therefore had a considerable impact on both the fishing industries and the economies of the South American Pacific countries. Decreased production in fisheries caused exports to fall, with corresponding reductions in foreign currency earnings.

Floods in Bolivia

Although many Bolivians are used to coping with perennial flooding, the severity of the 1982–83 rains caused exceptional disruption. The Mamore river burst its banks, flooding nearly 9 percent of Bolivia—about 98 000 km². The flooding affected more than 1200 homes, and more than 1000 families were evacuated. Some 38 000 cattle drowned, causing severe economic hardship in an area dependent on livestock.

One of the worst-affected towns was Santa Ana de Yacuma in the Beni district. The 1982 rainfall here was the highest for 50 years. In addition, the floods receded slowly, leaving pools of stagnant water that provided breeding places for mosquitoes and other disease-carriers. Cases of malaria in Guayaramerin, another town in the same district,

the effects of El Niño

Developing countries have few resources for dealing with emergencies such as flooding, and conditions rapidly deteriorated in Ecuador and Bolivia during the 1982-83 El Niño floods.

increased substantially during the second half of 1982.

Food shortages meant that people went hungry or ate rotting food. Cases of acute diarrhoea and respiratory infection doubled during April and May 1982 compared to 1981 levels.

A survey of small mammals showed that rats and other animals had moved closer to areas of human habitation to escape the water and to hunt for food, though fortunately no epidemic broke out.

The flooding may also have had some beneficial effects on health. After the flooding, the number of *Calumys callosus*—the rodent that carries Bolivian haemorrhagic fever—dropped compared to previous years. As *Calumys* is a large, cumbersome animal and unable to swim, many must have drowned.

Floods in Ecuador

The rainfall in Ecuador during the 1982–83 El Niño was so torrential that 'areas that normally measure precipitation in inches have had 10 feet', according to the director of emergency relief in the town of Guayaquil. Banana and rice crops rotted, 800 coastal houses were destroyed, and many coast roads and bridges were washed away. The final cost of flood damage to crops and property was estimated at $400 million.

Livestock suffered from foot-rot caused by the prolonged immersion, while poultry died from a fungal disease that spread quickly in the humid conditions.

In the shanty towns surrounding Guayaquil, sewage collected in stagnant pools of water that grew crusts of green slime, and typhoid reached epidemic proportions in the poorest areas.

The El Niño phenomenon

Floods in Peru

Countries along the Pacific coast of South America are often affected by severe weather during El Niño years, particularly torrential rain and high winds that lead to flooding and land slides. As well as disrupting transport, communications, agriculture and industry, these weather anomalies are a direct threat to life.

During the 1982–83 El Niño, two normally dry northern Peruvian regions, Piura and Tumbes, experienced heavy rain for nearly six months. In some regions there had previously been no rain for 10 years, and adobe buildings literally melted away in the downpour. Roads built on sand were eroded away, and water, electricity and drainage systems broke down. A state of emergency was declared.

A survey on health during the dry, first six months of 1982, compared with the wet conditions in 1983, showed that death rates from all causes increased by more than 90 percent in 1983. Increases in respiratory and gastrointestinal diseases were particularly sharp and the death rate doubled in the 1–4 year-old age group.

Malaria is endemic in the northern coastal areas of Peru. The flooding, combined with an increase in temperature and humidity, led to a sharp increase in mosquito populations. At the same time flood damage to property caused increased exposure of the human population to bites. The result was a malaria epidemic.

On the other hand, the incidence of malaria actually declined in the inland Andean valleys, and remained stable in the jungle areas, indicating that the dramatic increase was confined to the regions affected by El Niño rains and flooding. In the coastal region, the increase was 191

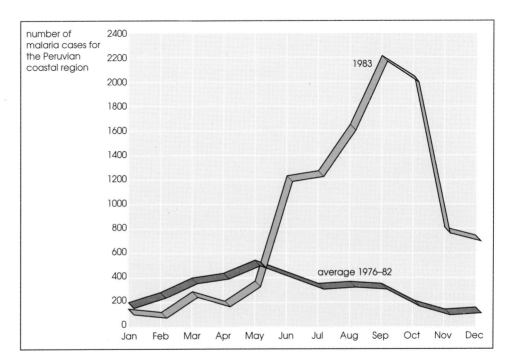

Figure 8 Incidence of malaria increased dramatically during 1983 in the coastal areas of Peru affected by El Niño flooding.

cases per 100 000 population. Malaria incidence increased most in the two regions, Piura and Tumbes, that were most severely affected by flooding (*see Figure 8*). In the latter, there were 1119 cases of malaria in 1983 compared with a yearly average of just 20 during the previous 6 years.

Heavy floods are common in many parts of South America, though they are not usually followed by such increases in endemic illness. The Trinidad and Beni provinces of Bolivia often suffer flooding, yet no epidemics were reported there even after the heavy 1982–83 floods. It may be that where floods are a recurrent problem, communities adapt.

In northern Peru, however, only the very old could remember rains as heavy as those in 1982–83, and it was the very abnormality of the weather conditions that exacerbated the death rate. For example, a sandy ravine crossed the centre of the town of Sullana. Some could remember water flowing down it 70 years previously, but in the intervening period, shops, houses and a market had been built along the water way. All these buildings were destroyed or inundated during the 1982–83 floods.

Intermittent, severe El Niño-induced flooding is probably a long-standing destructive force in Peru. Michael Mosely of the Chicago Field Museum of Natural History believes that sophisticated irrigation systems on the Peruvian plains may have been destroyed by El Niño rains as early as AD 1100. Without the irrigation systems, the area gradually became arid, as it is today.

Brazil

El Niño events have been connected to two contrasting climatic disturbances in different parts of Brazil. The north-east region, which is subject to periodic drought and variable rainfall, suffered extreme drought in 1983. In contrast, southern Brazil, which normally has high rainfall, experienced heavy and prolonged rainfall in 1982–83 resulting in extensive flooding.

The two regions also differ greatly in their level of development. Southern Brazil has modern agriculture, is heavily industrialized, and has higher and more evenly distributed incomes. The north-east has many more poorly paid subsistence farmers; incomes are generally lower and are unevenly distributed. This difference affects the ability of people in the two regions to adapt to climatic stress.

In 1983, about 88 percent of the north-east region—including 14 million people—was affected by drought. The drought caused a 16 percent decrease in agricultural production and many subsistence farmers lost all their production. Some food prices increased by 300 percent, agricultural unemployment soared, and the government had to give drought assistance to some 2.8 million people.

During the same year, in southern Brazil, rainfall increased by 70–100 percent. The area experienced the worst flooding of the century, with destruction of buildings, transport and communications networks. Scores of people died, and several hundred thousand were made homeless.

In the southern province of Santa Caterina alone, crop losses were estimated at $924 million. Losses of livestock amounted to 800 000 poultry, 6500 swine and 43 000 cattle. Water is estimated to have removed about 25 million tonnes of fertile topsoil. Despite these huge losses, and a drop in per caput income, the generally high level of income in the area enabled the population to subsist independently, in contrast to the north-east region.

Nationally, this extreme weather reduced harvests of export crops, such as coffee and soya, and exacerbated Brazil's huge trade deficit. The combined losses in production in the dry north-east and inundated south have been estimated at $875 million for the first 6 months of 1983—about 10 percent of the production expected for that year.

Cyclones over Polynesia

El Niño events affect cyclone activity in the Pacific region. The centre of cyclone activity is normally in the western Pacific but, during an El Niño, this area moves to the east where cyclones become unusually prevalent.

Six major cyclones struck French Polynesia from December 1982 to April 1983. On Tahiti, 1500 houses were flattened and 6000 lost their roofs. Storm warnings are thought to have helped save many lives—only 14 Tuamotuans died compared to 500 in a 1903 hurricane—but many outer islanders migrated to the central islands after the El Niño-induced storms. There is anxiety that such migration into the urbanized central islands will accelerate the destruction of the traditional Polynesian cultures.

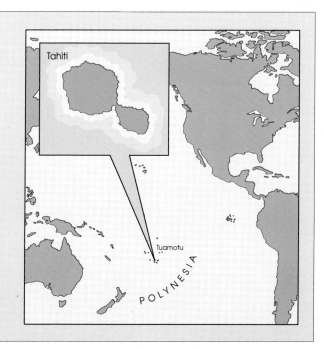

the effects of El Niño

Wildlife disappears from Pacific Islands

The Galápagos Islands lie in the path of El Niño currents in the eastern Pacific. The 1982–83 El Niño had profound effects on the wildlife of the islands, and hence on the islands' tourist industry. As warm El Niño currents reduced plankton productivity, many marine creatures starved, as did the land animals and birds which fed on them. A marine biologist stationed on the Galápagos commented, 'Stories of marine iguanas dying at visitors' feet, of fewer or no sea birds along park trails, and starving fur seals circulated in the travel business. The problem was, the stories were true.'

Nine times as much rain fell as normal (*see Figure 9*). The normally arid Galápagos Islands were transformed into lush tropical islands. Roads became quagmires of mud, and were often interrupted by rushing rivers. The floor of several volcanic craters became giant cisterns of fresh water.

Waved albatrosses, which breed almost exclusively on the Galápagos islands, failed to hatch any eggs in 1983. Rain had encouraged a thick coat of vegetation to cover their normal nesting grounds, and their fish diet was in short supply. Fortunately one good reproductive year can make up for disastrous years like 1983.

The same is not true of seals and sea lions. During 1983, food shortages resulted in the death of almost all seals and sea lions

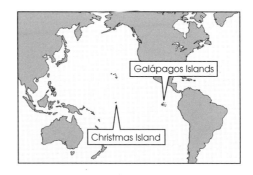

under five years old, and of about 30 percent of the adult seal population.

Blue-footed booby birds also apparently vanished from the Galápagos, although once the sea temperatures began to fall, some returned to their previous haunts.

Food shortages also affected the marine iguanas that normally feed on leafy algae along the intertidal zones. During the El Niño these algae were replaced by a different, less nutritious type of algae. Many iguanas starved and nearly 90 percent of the 1982 iguana hatchlings died.

On Christmas Island in the Pacific, the effects on wildlife were just as dramatic. In November 1982, the normal population of several million sea birds had all but vanished, leaving nestlings to starve. Unusual sea temperatures and currents had severely reduced the stock of fish on which the birds' survival depends.

Figure 9
Sea-surface temperatures around the Galápagos Islands rose during the 1982–83 El Niño and rainfall increased by nearly 10 times. Effects are also intense on Christmas Island, which lies in the path of El Niño (see above).

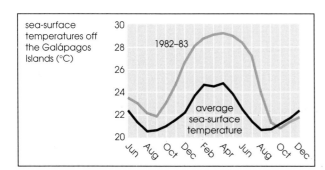

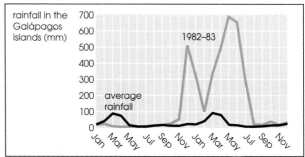

The El Niño phenomenon

Drought in Australia

Figure 10 shows decreases that occurred in a number of important sectors of the Australian economy during the El Niño years of 1982–83.

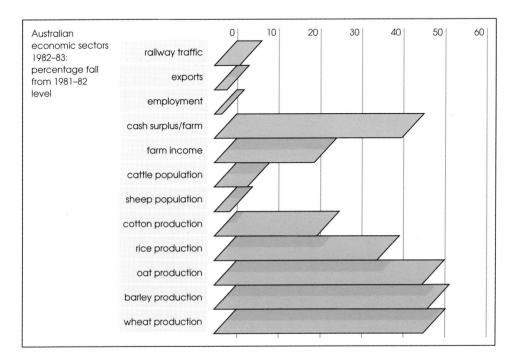

Australia has suffered severe droughts in its northern and eastern regions during many El Niño years. These droughts have reduced crop yields, killed livestock, eroded soils and encouraged destructive bush fires. The El Niño event of 1982–83 coincided with a drought that may have been the worst for 100 years. The 1982 winter rains (June-September) failed over the south-eastern grain and pasture areas of Australia. In northern New South Wales and southern Queensland, the summer rains (December 1982 to February 1983) also failed.

The drought was broken in the autumn (March to May) of 1983 by torrential rains in eastern Australia. Sheep, weakened by the drought, suffered severely: many were drowned or starved, and foot-rot caused many lengthier deaths.

The drought exacerbated land degradation. Powder-dry soils blew away in the wind. A dust storm on 8–9 February 1983 blew away an estimated 150 000 tonnes of soil from farms in north-west Victoria. More than 11 000 tonnes of dust landed on Melbourne; some was blown as far as New Zealand.

Bush fires killed 72 people and left 8000 homeless. They caused the deaths of about 300 000 animals, and property damage estimated at A$400 million. When the rains finally came, the burned areas were severely eroded, some having lost as much as 43 tonnes of soil per hectare.

Wildlife losses were not well documented but estimates suggested that up to 70 percent of the kangaroo population in commercial shooting zones died between November 1982 to June 1983.

The total cost of the 1982–82 El Niño in Australia was estimated at more than US$2500 million.

Drought and forest fires in Indonesia

Indonesia suffered a drought during the 1982–83 El Niño that affected the country's rice crop and contributed to vast forest fires in the tropical Kalimantan region.

Evidence suggests that Indonesian droughts are strongly associated with El Niño events. An analysis of sea-salt production—which is closely linked to weather conditions—in Madura, Java, over many years indicates that more than 90 percent of drought periods have occurred during an El Niño event; other records have associated 78 percent of east monsoon (April to October) droughts with El Niño.

Since the early 1970s, Indonesia has introduced irrigation, and new seed varieties, fertilizers and pesticides, and has progressed from being the largest importer of rice, to self-sufficiency in this product.

However, the 1982–83 drought severely retarded this rapid expansion of rice production. There are normally two rice crops in Indonesia, one during the dry season (May to October) and a larger one in the wet season (November to April). During the 1982 dry season, rainfall was very low. In the main rice-producing regions there was no significant rainfall for four to five months, and the start of the wet season was delayed for about a month. The drought affected the dry-season crop in 1982 and delayed the planting of the 1982–83 wet-season crop.

In certain areas, this led to local food shortages, disruption of the drinking-water supply, loss of cattle, and to outbreaks of cholera. Throughout Indonesia, the crop failure of 1982–83 led to more than 300 deaths.

The death toll could have been much higher but for many farmers switching to maize (a more drought-resistant crop) after the first signs of the drought. Maize production increased by 56 percent in 1983, and partly offset the fall in the 1983 wet-season rice crop.

Despite its description as the worst drought of the decade, rice production did not drop below 1980 levels, and by 1984 the situation appeared to have returned to normal. Indonesia's intensive agricultural system, irrigation and crop substitution helped reduce the effects of the drought.

Forest fires in East Kalimantan during 1982–83 were in many ways much more serious. More than 3.5 million hectares in East Kalimantan, and 1 million hectares in Malaysian Sabah were damaged. Fires raged for almost three months and were described as one of the worst environmental disasters of the century.

The average annual rainfall in the forests is more than 2000 mm, enough to make the area almost invulnerable to drought. However, lack of rain in 1982–83 caused normally evergreen trees to lose their leaves, which resulted in a build-up of dry litter on the forest floor. Fires may have been triggered by agricultural burning for land clearance and, once kindled, they spread rapidly through the litter. Forest containing areas of logging suffered badly because organic debris from the logging was left on the forest floor.

Smoke from the fires closed airports and ports, and reached as far as the Malaysian peninsular, some 1500 km to the west. Peat swamp forest that had dried out to expose peat surfaces burned in great, slow underground fires.

Though Indonesia's intensively developed agricultural system recovered fairly quickly from the El Niño drought, the tropical forest of Kalimantan and its delicate ecosystem will take much longer to return to their former condition.

> More than 3.5 million hectares in East Kalimantan, and 1 million hectares in Malaysian Sabah were damaged. Fires raged for almost three months and were described as one of the worst environmental disasters of the century.

The El Niño phenomenon

El Niño and Indian Monsoons

During 1982–83, more than one-third of India was affected by drought

More than half of India's gross national product comes from agriculture, the success of which depends on the seasonal monsoon rains. When the monsoon is erratic, it can cause droughts in some parts of the country and floods in others.

Despite intensive study, no reliable method of predicting monsoon rainfall has yet been found. However, many of the years when the monsoon has been weak, and there has been widespread drought, have also been El Niño years.

During 1982–83, more than one-third of India was affected by drought. The monsoon began late, then in many places brought 150–250 mm of rain within a single 24-hour period, and finished early. These droughts and deluges disrupted agricultural operations, and caused a 3.7 percent decline in agricultural production. This figure hides disastrous effects in particular regions, where agricultural production fell by as much as 50 percent.

However, total grain production was relatively unaffected and fell by only 5 million tonnes. This compares favourably with the drought of 1978–79 (a non-El Niño year), which caused a water deficit over nearly half the country, and led to a drop in grain production of more than 22 million tonnes.

The explanation for this lies in deliberate efforts to adapt grain production to climatic conditions. Indian grain production occurs in two main seasons. The *kharif* (monsoon) season is hot, and production levels depend largely on the monsoon rainfall. The *rabi* (winter) season is cooler, and the crops grown require less moisture. When the monsoon rains are weak, and production falls, efforts are made to increase winter season production. Over the past 25 years, winter season production has been increasing and the difference in production between the two seasons has narrowed.

This policy has helped to cushion food production from the effects of weak monsoons. In 1982–83, the kharif crops were reduced by about 10 million tonnes, but winter production increased by about 5 million tonnes, partly compensating for the loss.

To further enhance food security, the Indian government has, since the 1970s, revived the ancient custom of holding reserve stocks of grain to distribute during times of food shortage. In 1982–83, these stocks amounted to 18 million tonnes, and the government was able to intervene to supply food for the needy at reasonable prices. Without government intervention, the poor would have found it difficult to afford many basic foods.

Figure 11 shows percentage variations from the average monsoon rainfall over India. El Niño years, shown in colour, are regularly years of severe drought.

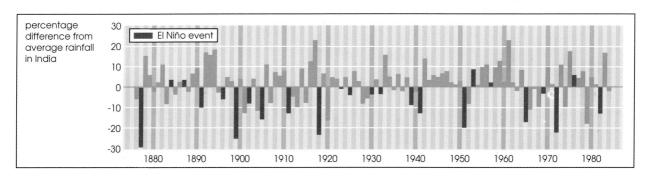

the effects of El Niño

El Niño in China

Connections between El Niño events and climatic anomalies are not as dramatic or as consistent for China as they are for Australia, Indonesia and Latin America. Weather conditions in 1982–83 were unusual in certain areas of China, and crop yields were affected. However, national levels of crop production were not greatly reduced, partly because policy and technology changes have increased yields dramatically and this in some ways obscures the effects of climatic anomalies, and partly because of the wide variety of climates found in this vast country.

As in Indonesia, crop production increased rapidly in China during the 1970s and 1980s. However, adverse weather during 1981–82 led to a fall in production in north-east and central China. The winter of 1981–82 was drier than usual in the North China Plain area in the north-east, and grain production consequently fell by about 10 percent. The summer of 1982 was cold and wet in central China, with severe flooding in the area between the Huangho (Yellow) and Changjiang (Yangtze) rivers, where production of soybeans, beets, peanuts sugar cane and cotton fell.

It is difficult to link these effects directly with El Niño, as the 1982–83 event developed late—sea temperatures off Peru did not become abnormally high until August. But of 30 cool summers during the period 1860–1980 in north-eastern China, 16 were during El Niño events, and 12 occurred within the year preceding or following an El Niño year.

Climatic anomalies during 1983 were much more pronounced, and there is more meteorological evidence to link these to El Niño. During the winter of 1982–83, rainfall over south China exceeded the average by between 50 and 300 percent,

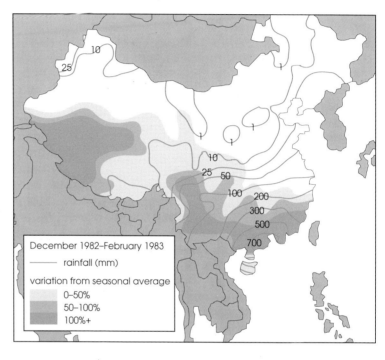

leading to floods. Before this, winter flooding in this area had not been observed, according to records going back to the beginning of the century. The wet trend persisted until April, damaging early wheat and rice crops. In some south China provinces, wheat yield was 60 percent below that of the previous year. Fortunately, good autumn weather provided record rice harvests, compensating for the low spring yields.

Although the 1983 winter floods were apparently associated with El Niño, other El Niño years have resulted in inconsistent winter weather anomalies in China—sometimes flooding and at other times, droughts. The extreme flooding in 1982–83 probably resulted from the unusual intensity of the El Niño, whereas in other years El Niño is not strong enough in China to affect the weather consistently.

Figure 12 shows the exceptionally heavy rainfall during the winter of 1982–83 (more than three times heavier than normal in some areas) which caused severe flooding in southern China.

The El Niño phenomenon

Drought in southern and eastern Africa

The countries of southern and eastern Africa have very different climates, but most are situated in semi-arid regions where annual rainfall varies by 20–30 percent. This variation can have devastating social and economic consequences. In most of the area, agricultural production is mainly limited by the availability of soil moisture. Heavy rains often do not significantly increase soil moisture because the soil does not absorb moisture quickly, and the rain tends to run off without penetrating the soil. In addition, intense sunlight and high temperatures cause rapid evaporation of moisture from the soil. The problems are aggravated by the tendency of 2–3-year droughts to occur at fairly frequent intervals.

Evidence for a strong link between El Niño events and African drought is conflicting. Over the period 1875–1975, 27 El Niño events were linked to 21 years of deficient rainfall in southern and eastern Africa. The 1982–83 El Niño was also linked to drought in Africa. More detailed attempts to link the two phenomena have not been successful.

In southern Africa, agricultural production in Zimbabwe and Mozambique was reduced by the drought in 1982–83. Although 1982 was a good year for Zimbabwe, there were heavy rains early in the 1983 season, followed by drought. The 1983 harvest was down 65 percent on 1982. Money was diverted from development projects into emergency drought relief. The south and west of the country lost livestock as well as crops. Transport and agro-industries were affected by the drop in production, and water was rationed.

Mozambique's 1982–83 drought was considered the worst in 50 years and led to

Devastating droughts in southern and eastern Africa in 1982–83 have been linked to El Niño.

many deaths. There was an exodus of refugees to neighbouring Zimbabwe, also hard hit by drought.

In Botswana, cattle outnumber the human population by three to one. During 1982–83, key watering places dried up completely and livestock mortality was high. Almost half the population was fed by emergency relief from overseas. Normally one child in four in Botswana is considered at risk from malnutrition, but the figure increased during the drought to one in three.

In eastern Africa, Ethiopia and Tanzania had good years in 1982, but production was stagnant in 1983 and fell in 1984. Kenya and Uganda showed steady growth throughout the period. Kenya even resumed food exports in 1983, with an excellent tea crop, although grain production was down by about 15 percent on 1982.

During drought years, which often coincide with El Niño years, the need for emergency food imports can deplete scarce foreign currency reserves. Climate can then have indirect effects on development projects as governments cut back on expenditure to compensate for the foreign exchange losses. Foreign exchange shortages also often lead to scarcities of agricultural inputs such as fertilizers and pesticides. This prolongs the effect of the drought year, affecting future crop yields.

However, international market prices can sometimes offset drought-related crop losses. In 1984, the main March-May rains failed in many parts of eastern Africa, causing a severe drop in agricultural production (1984 was not an El Niño year). Kenyan tea and coffee export prices more than doubled as a result of scarcity. This resulted in an increase in foreign exchange earnings despite the drought.

Floods in Thailand

Bangkok had torrential rainfall during 1983. The rain was dubbed 'the thousand-year rainfall' and flooded parts of the city for almost four months.

In addition to the direct damage caused by the flooding, the rains caused substantial erosion, which affected both agriculture and fishing. The record rainfall washed minerals and organic material into the gulf of Thailand. This resulted in a sediment that Thai fisherman called 'whale shit'. This sediment had a detrimental effect on plankton, which in turn affected the fish population. The oyster industry was also badly hit, as oysters failed to breed, and died in great numbers.

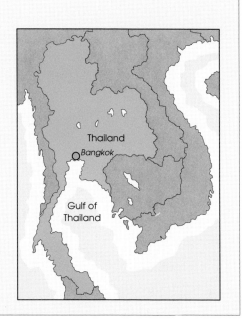

The El Niño phenomenon

Storms over North America

During the 1982–83 El Niño, the weather was highly unusual in many areas of North America. Perhaps the most dramatic event was Hurricane Iwa, which struck the Hawaiian Islands in November 1982 (the first hurricane for 23 years) and caused property damage estimated at $234 million. Sugar cane crops and orchards were also damaged. El Niño pressure anomalies may have caused an eastward shift in the area where tropical Pacific storms are formed, leading to the direct strike on Hawaii.

During the winter of 1982–83, the west coast of the United States also suffered severe storms. High winds, coastal erosion and flooding caused damage amounting to several million dollars. Farmland was flooded; roads damaged; and grain, vegetable and fruit crops destroyed. Nearly 10 000 homes were affected by mud slides and 1000 businesses were damaged or destroyed. One positive effect of the weather was a sharp rise in the water table. As half California's water is supplied from wells, the inundation provided insurance against future drought. The El Niño was probably responsible for driving severe weather into California during 1982–83 yet, during the 1976–77 El Niño, instead of suffering torrential rain, California was hit by severe drought.

After devastating California, the rain-bearing clouds moved eastwards and there was a dramatic increase in the snowfall on the Rockies in the winter of 1982. When this melted, it resulted in severe spring flooding in Salt Lake City and in many states of the lower Colorado river valley, particularly Nevada, Colorado, Arizona and New Mexico.

The 1982–83 El Niño also brought

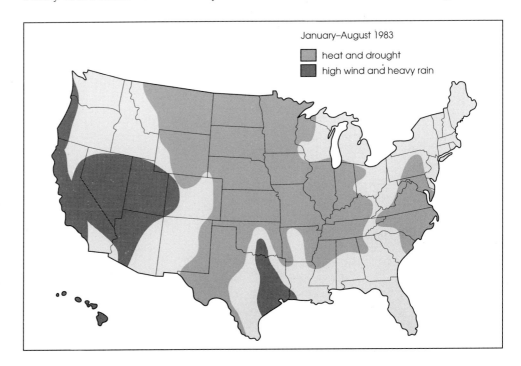

Figure 13 The winter of 1982–83 saw rain and flooding in the west and south of the United States, and the summer brought heat and drought to the corn belt and Great Plains region.

torrential rainfall and flooding along the Gulf Coast states and in the Midwest. Here, overall damage amounted to more than $1000 million. Fifty people died and thousands were evacuated in the states of Louisiana, Mississippi, Arkansas, Missouri, Illinois and Iowa. The downpours caused severe soil erosion and damage to forage, as well as to roads, bridges and farm buildings. During April 1983, more than 100 000 poultry and 1000 cattle died. In May, 17 percent of the cotton crop was lost.

The Mississippi valley was badly affected, with almost 100 000 homes flooded and many roads blocked. Potentially disastrous Mississippi flooding was narrowly averted by sandbagging. In New Orleans, almost three-quarters of the city's streets were submerged under 1.5 metres of water.

On the positive side, the eastern United States had a warm winter with temperatures 3–6 °C above normal. Energy demands for heating were reduced by about $2500 million, stress on livestock was reduced and local government saved money on snow removal. The mild weather had adverse effects on ski resorts and utility companies, on the other hand.

Summer 1983 saw a drought and heat wave affecting the corn belt and Great Plains areas of the United States. The warm weather began in mid-July 1983, affecting developing crops of corn, soybeans, sorghum and cotton, and causing significant yield reductions. These were estimated at $5500 million for corn, $480 million for cotton, $500 million for sorghum, $1100 million for tobacco and $3.4 million for soybeans.

The El Niño event was also blamed for disrupting fishing off the north-west coast of the United States. Salmon catches off Washington State were halved and competition for the fish caused great conflict amongst fishermen. A treaty between Canadians and Americans for equal sharing of sockeye salmon was circumvented by the fish, which chose another route around Vancouver Island to return to the Fraser River, leading to greatly increased Canadian catches, and severe reductions for US fishermen. North-west coast fishing incomes decreased by 10 percent during 1983, and the cost to the fishing industry of the decline in catches was estimated at $400 million.

Other fish were also displaced from their northern grounds. Anchovies, bonito, barracuda, red crab, popeye catalufa and bocassio were found scattered to the north of their traditional grounds. The distortions in normal fishing populations also led to disruptions in bird habitats. Puffins, cormorants and gulls declined in numbers, whilst terns, brown pelican and frigate birds migrated to new nesting grounds. The warmer sea temperatures had positive effects on some other fish populations, with albacore tuna, oysters and yellowfin tuna all increasing.

Implications for policy

The devastating loss of life, damage to property and infrastructure, and disruption of agricultural production caused by an El Niño can be minimized through national and international planning and policies. That governments have not implemented such policies widely to date is partly due to incomplete understanding of El Niño and inability to predict its effects reliably beyond the Pacific region. In countries where an El Niño and its effects can be predicted with relative certainty, governments are often unable—as a result of lack of resources or political continuity—to develop and implement long-term strategies to minimize the effects of El Niño.

Climatic anomalies rarely confine themselves within national boundaries, and the evidence for El Niño's teleconnections underlines the international nature of the problem. For this reason, international organizations have an important role to play in coordinating international research projects and information exchange on El Niño, raising international awareness of the phenomenon and helping nations develop individual or combined response strategies.

As part of the World Climate Programme jointly implemented by several United Nations bodies and the International Council of Scientific Unions, UNEP implements the World Climate Impact Assessment and Response Strategies Programme (WCIRP). As part of this programme, UNEP organized a workshop in 1985, held in Lugano, Switzerland, on the socio-economic impacts of the 1982–83 ENSO event. Social and physical scientists presented evidence for proposed teleconnections around the world and the effects that the 1982–83 ENSO had in a variety of countries. Other workshops were held in Bangkok in 1988 and in 1991 to discuss new research findings, including the issue of climate change and El Niño, and establish plans for further research.

Research to date on the occurrence of El Niño, its effects on global climate anomalies and its impact on different societies and forms of agriculture can furnish governments with basic information. However, countries also need to carry out their own national or multinational climatic impact assessments to predict the likely effects of El Niño events on natural resources, and to assess their social and economic consequences. Such studies would be most useful in countries that experience consistent effects from El Niño, and would enable governments to devise policies that would effectively minimize its impact.

Impact assessments focus on the economic sectors most susceptible to disruption by weather anomalies—the most important of these are likely to be agriculture, water resources, marine fisheries and forests. The extent of the disruption which climatic anomalies can cause within a community or country depends on existing social and economic conditions. Therefore, distribution of income, type of agriculture, availability of new agricultural land and secondary sources of income, population distribution, type of housing, the provision of public amenities, and so on, also need to be taken into account in any impact assessment.

Once impact has been assessed, policies can be devised to minimize the effects of El Niño. In countries far from the Pacific, policies are likely to be short term, and only implemented for the duration of El

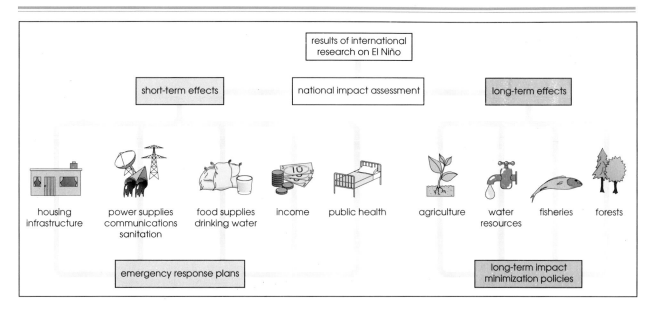

Niño conditions. In other countries, adapting to El Niño may necessitate expensive building programmes or permanent changes in agricultural practices.

For instance, meteorologists in Ethiopia studied US data on El Niño events from the 1950s to assess how El Niño would affect their own weather conditions in 1986–87. On the basis of a prediction of good spring rains and a failure of the summer rains, the government issued seeds, fertilizer and other agricultural assistance to farmers in time for the spring planting, and production increased by one-third. The summer rains did fail, and the increased spring yield helped to cushion the impact of the reduced summer harvest.

Countries nearer the Pacific need to be even more adaptable, if El Niño is not to cause social and economic devastation. As well as causing hardship to fishermen, the collapse of the Peruvian fisheries in the 1970s meant that Peru could no longer export the same quantity of fishmeal to the United States. The price of fishmeal increased in the States due to scarcity, demand fell, and US farmers switched to domestically produced soymeal, which then rose in price. Having lost the fishmeal market entirely, Peru began to change from growing corn and wheat to producing soymeal for export to the States, and its production level has now increased beyond that of the United States.

Being able to minimize the effects of El Niño—or even to take advantage of them—depends not only on reliable forecasts of El Niño events, but also on agreement between scientists and policy makers as to their likely effects and effective counter-measures. Without a consensus of opinion, policies are unlikely to be formed, thereby reducing the value of El Niño forecasts.

For instance, in 1977, experts from government, fisheries, and public policy makers were asked what action should have been taken in 1972 to save the Peruvian fisheries. Their replies ranged from imposing a ban on all fishing to protect the shoals during El Niño, to allowing all-out fishing to thin the shoals and improve their chance of survival.

Even if accurate forecasts were available, it is not only scientific information that decision makers taken into account: economic and political constraints also come into play. For instance, even if a forecast had been given about the 1977 El Niño, cautious exploitation of fish stocks could have been rejected in favour of plans for national economic expansion or for other political reasons.

In the face of a problem of international proportions, the way forward lies in cooperation between nations in funding research and pooling information on El Niño and in devising national and multinational projects to reduce its effects.

Figure 14
Information on El Niño can be used as the basis for national impact assessments and from these governments can draw up emergency relief plans and long-term strategies to minimize the effects of El Niño.

Sources

Canby, T. Y. 'El Niño's ill wind'. In *National Geographic*, February 1984.

Chen, R. S., and Parry M. L. (eds.). *Climate impacts and public policy.* Nairobi, UNEP, 1987.

Glantz, M. H. 'El Niño: lessons for coastal fisheries in Africa?' In *Oceanus,* Summer 1980.

Glantz, M. H. 'Floods, fires and famine: is El Niño to blame?' In *Oceanus*, Summer 1984.

Glantz, M., Katz, R., and Krenz, M. (eds.). *Climate Crisis.* Nairobi, UNEP, 1987.

The Global Climate System (Autumn 1984–Spring 1986). Geneva, WMO/UNEP/GEMS, 1987.

González, C., Gueri, M., and Morin, V. 'The effect of the floods caused by El Niño on health'. In *Disasters*, 10/2/1986.

IPCC, *Climate Change: the IPCC impacts assessment.* WMO/UNEP, 1990.

Lundine R., Motha, R., and Puterbaugh, T. 'An ill wind: El Niño rainfall anomalies and regional crop yield variability'. In *Mazingira* 8 (6), 1985.

Nicholls, N. 'Towards the prediction of major Australian droughts'. In *Australian Meteorological Magazine* 33 (4), December 1985.

Rasmusson, E. M. 'El Niño variations in climate'. In *American Scientist* 73, March–April 1985.

Robinson, G. R. 'Negative effects of the 1982–83 El Niño on Galápagos marine life'. In *Oceanus,* Summer 1987.

Russac, P. A. 'Epidemiological surveillance: malaria epidemic following the Niño phenomenon'. In *Disasters* 10/2/1986.

Stuller, J. 'El Niño stirs up the world's weather'. In *Reader's Digest*, January 1984.

Telleria, A. V. 'Health consequences of the floods in Bolivia in 1982'. In *Disasters,* 10/2/1986.